# Table of Contents

**INTRODUCTION**

**MARKOV MODELS**

**CHAPTER 1: MARKOV MODELS**

    AXIOMS TO UNDERSTAND MARKOV MODELS
    THE MARKOV PROPERTY
    MARKOV MODELS AND OBSERVATION PROBABILITIES
    HIDDEN MARKOV MODEL

**CHAPTER 2: DATA MINING**

    SOFTWARE
    STRATEGIES

**MACHINE LEARNING**

**CHAPTER 3: WHAT IS MACHINE LEARNING?**

    SUBJECTS INVOLVED IN MACHINE LEARNING
    VARIETIES OF MACHINE LEARNING
    USES OF MACHINE LEARNING
    APPLICATIONS OF MACHINE LEARNING IN THE REAL WORLD

**CHAPTER 4: SUPERVISED MACHINE LEARNING**

    OVERVIEW
    ISSUES TO CONSIDER IN SUPERVISED LEARNING:

**CHAPTER 5: UNSUPERVISED MACHINE LEARNING**

**ARTIFICIAL INTELLIGENCE AND DATA SCIENCE**

**CHAPTER 6: ARTIFICIAL INTELLIGENCE 101**

    INCREASED COMPUTATIONAL RESOURCES

- Growth of Data
- Deeper Focus
- Knowledge Engineering
- Alternative Reasoning Models
- Exploring AI

## CHAPTER 7: BIG DATA AND ARTIFICIAL INTELLIGENCE

- What is Big Data?
- Assessing Data using AI

## CHAPTER 8: HOW DOES AI DRAW CONCLUSIONS FROM DATA?

- Deductive Reasoning
- Inductive Reasoning
- Abductive Reasoning
- Case Based Reasoning
- Common Sense Reasoning

## CHAPTER 9: WHAT IS DATA SCIENCE?

## PYTHON

## CHAPTER 10: MARKOV'S MODEL IN PYTHON

- Getting Started
- Programming a Hidden Markov Model

## CONCLUSION

# Introduction

Thank you for purchasing the book, *'Markov Models Supervised and Unsupervised Machine Learning: Mastering Data Science & Python.'*

In the following book, you will find a great deal of information on the mechanics of the Markov Model and how the model could be used in the world of machine learning though using the powerful yet simple language, Python. This book is for those who have just begun to study and understand Artificial Intelligence and Data Science.

Markov Models was discovered in the year 1916, by Andreevich Markov, a scientist who was studying and analyzing the frequency of different types of words in Pushkin's poems. Since then, the model created has been expanded to include representations of a number of probabilities like the Hidden

Markov Model that is integral to our understanding and towards building artificial intelligence.

The structure of this book will define each section of the topic piece by piece. We will first take a look at Markov Models and the mathematical aspect of those models. We will then take a look at Hidden Markov Models (HMM) and also study the three problems of HMM and also identify the solutions to the same. After thoroughly investigating the Markov Models, we will look at different aspects of Machine Learning and look at Supervised and Unsupervised Machine learning in greater detail.

After understanding the concepts of Markov Models and Machine Learning, we will take a look at Artificial Intelligence and Data Science and understand certain aspects of them. We will then learn about how to apply these concepts in Python. Everything from installing

Python to programming a Markov Model in Python will be covered in detail.

At the end of the book, you will have a clear understanding of the mathematical aspects of Markov Models and Hidden Markov Models. You will also be able to build enough confidence to help you program Markov Models using Python.

Thank you for purchasing the book. I hope you find it helpful.

# Markov Models

## Chapter 1: Markov Models

### Axioms to understand Markov Models

There are certain axioms that exist in probability and statistics that will need to be understood before looking at Markov Models.

***Fundamental Axioms***

Let us take a look at the various axioms that are used in Markov Models to understand the math that supports the model. One of the most fundamental axioms can be expressed as follows: $0 < P(A) < 1$

This states that the probability of any event occurring is always going to be greater than zero but less than one, both inclusive. This implies that the probability of the occurrence of

# Markov Models Supervised and Unsupervised Machine Learning:

*Mastering Data Science & Python*

© **Copyright 2017 by Healthy Pragmatic Solutions Inc. - All rights reserved.**

The contents of this book may not be reproduced, duplicated or transmitted without direct written permission from the author.

Under no circumstances will any legal responsibility or blame be held against the publisher for any reparation, damages, or monetary loss due to the information herein, either directly or indirectly.

**Legal Notice:**
You cannot amend, distribute, sell, use, quote or paraphrase any part or the content within this book without the consent of the author.

**Disclaimer Notice:**
Please note the information contained within this document is for educational purposes only. No warranties of any kind are expressed or implied. Readers acknowledge that the author is not engaging in the rendering of legal, financial, medical or professional advice. Please consult a licensed professional before attempting any techniques outlined in this book.

By reading this document, the reader agrees that under no circumstances are is the author responsible for any losses, direct or indirect, which are incurred as a result of the use of information contained within this document, including, but not limited to, —errors, omissions, or inaccuracies.

any event can never be negative. This makes sense since the probabilities can never be more certain than hundred percent and least certain than zero percent.

## *Additive Property*

The next axiom that is important to note can only be true when both events A and B are mutually exclusive.

$P(A+B) = P(A) + P(B)$

This axiom states that the probability of both events A and B occurring is the same as the sum of the individual probabilities of the events occurring if and only if both events exclude each other. For example, if A is the event that we get a six on rolling a die and B is the event that we get a five rolling a die, then this axiom holds. But, if B were assumed to be the event where we get an even number, this set would include the number six making the axiom false.

## *Joint Probability*

This property can be expressed as follows:

$P(a, b) = P(A=a, B=b)$

This property can be read as the probability of a and b is the same as the probability that event A turns out in state 'a' and event B turns out in state 'b'.

## *Conditional Distribution Axiom*

The conditional probabilities can be found using the following equation:

$P(X=a \mid Y=b) = [P(X=a, Y=b)] / [P(Y=b)]$

The equation above is read as, "The probability of the random variable X being a and the random variable Y being b is equal to the probability of X being a AND Y being b divided by the probability of Y being b".

The conditional distribution in statistics takes into account the probability that both events will occur given that one of the events has already occurred. The probabilities thus obtained are larger than the probability of the events X and Y occurring had Y not occurred earlier.

## *Chain Rule*

The chain rule is important for one to understand before taking a look at Markov's Model. This rule helps you obtain the joint probabilities of many variables with different properties. The equation can be expressed as follows:

$$P(a_1, a_2, a_3, ...., a_n) = P(a_1 \mid a_2, a_3, ...., a_n) * P(a_2 \mid a_1, a_3, ...., a_n) * .... P(a_{n-1} \mid a_n) * P(a_n)$$

This states that the probability of the occurrence of every event is equal to the probability that one event occurs given that every other event has already occurred multiplied throughout by the probability of the next event occurring given that the subsequent events have occurred and so on. This creates a chain of multipliers, which would need to be solved to obtain the probability for many events.

### *The Product Rule*

There is a product rule that is based on the independence of events, which states that:

$$P(A, B) = P(A \mid B) * P(B) = P(B \mid A) * P(A)$$

This states that the probability of the occurrence of both events A and B is given by the probability of the occurrence of A assuming

B has already followed, and multiplied with the probability of B. This is also equal to the probability of the occurrence of the event B given that A has already occurred multiplied by the probability of A. If you look closely at this equation, you will realize that this axiom simply multiplies the conditional distribution axiom with P (B) on both sides.

### *Independence of Multiple Probabilities*

The independence of multiple probabilities can be confirmed using the following axiom:

P (X) = P (X | Y)

In simpler words, the probability of the event X occurring is equal to the probability of X occurring given that the event Y has already occurred. If this axiom were true, it would imply that the occurrence of the event Y does not affect the occurrence of event X because the probability of X is unchanged irrespective of what the probabilities of Y are.

### *Bayes' Rule*

This rule is often used to identify the conditional probability when P (A, B) is not known. The equation used is as follows:

P (A|B) = [P (B|A)*P (A)]/P (B)

These various axioms are a great way to understand the logic behind Markov Models. Let us now take a look at the mathematical aspects of the Markov Model. As mentioned above, Markov Models was discovered in the year 1916 by Andreevich Markov, a scientist who was studying and analyzing the frequency of different types of words in Pushkin's poems. These models have now become an integral model to use while working with data science, artificial intelligence, and machine learning.

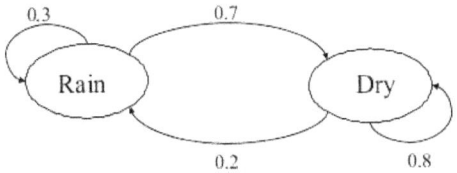

- Two states : 'Rain' and 'Dry'.
- Transition probabilities: P('Rain'|'Rain')=0.3 , P('Dry'|'Rain')=0.7 , P('Ra')=0.6 .
- in'|'Dry')=0.2, P('Dry'|'Dry')=0.8
- Initial probabilities: say P('Rain')=0.4 , P('Dry

## The Markov Property

In any system, the state of the system is constantly changing. This is the premise on which the Markov Models were developed. For any system to be a Markov model, it will also need to adhere to the Markov Property.

The property states that the past does not affect the future of any system, i.e. the next state of the system depends only on the current state of the system and only the single previous state affects the next state. For instance, if you were to consider the weather as a Markovian system, you can say for certain that if the weather was sunny yesterday and is humid today, the probability of it raining tomorrow would only depend on today.

# Markov Models and Observation Probabilities

Before building a Markovian system, it is important to know the different states of the system. Let us consider an employee who is very sick. There are three states that he could be in – alive and healthy, sick and dead. We will then need to arrange all these states with their transition probabilities (probability of moving from one state to the next) in a matrix. The probabilities of moving from one state to another are identified and are calculated as the probability of being in one state at a particular time t+1.

Finally, it is important to understand the difference between state and observation. At times, you will be unable to tell what the weather is like outside when you are at work. You will guess by what people are wearing. The observations that you make are what form the observation set. These observations are very different from the states since the states are definitive while observations are not.

## Hidden Markov Model

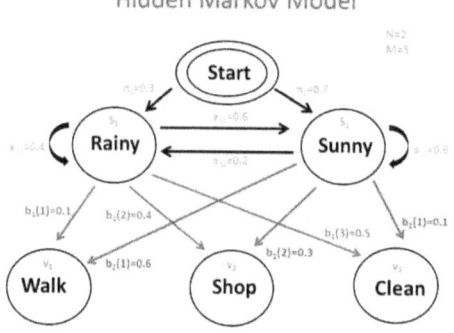

The observational probability distribution that is obtained set of observations made is an integral part of the Hidden Markov Model. In a given Hidden Markov Model, the actual states are unknown to the observer implying that the observer can only make any observation of the effect of a given state. We have looked at an example of a person who is inside trying to make a decision on the weather outside. This could be the case in many circumstances. Hidden Markov Models are quintessential to machine learning for this reason. The variable that is in question is often not measurable or observable directly but can be measured by the effects it has on its surrounding environment. The probability of an observed state is called the emission probability.

Now let us take a look at the problems that the Markov Models and Hidden Markov Models cause – evaluation problem, decoding problem and learning problem. The evaluation problem looks at the likelihood of a sequence of observations considered, the decoding problem looks at the different states behind the observations made so far and the learning problem, the most complex problem determines the likely values for the events that may occur.

# Chapter 2: Data Mining

Data mining is a process that is used by numerous businesses to convert the collected raw data into information that can be used by the business. There are specialized tools that can be used to detect patterns in large – scale information which would help you learn more about your consumers and also respond to their concerns while developing strategies that would help to increase your revenue. The ultimate goal of data mining is to increase your profits. Data mining is said to be effective only when data is gathered and stored in an efficient manner and processed in the right manner for future use.

Retail shops, like supermarkets and grocery stores, use data mining the most when compared with larger firms. There are a number of retailers offering loyalty rewards, which would motivate their customers to purchase only from that retail store since they would be able to reduce their expenditure. These rewards make it easier for the businesses to monitor who is purchasing what product, the time they purchase that product and also their usual budget. After conducting a thorough analysis, the retail store could choose to use the information obtained for different purposes, like providing customer coupons specifically targeted to their purchasing intent and deciding when to place items on sale or when to sell specific products at a higher markup.

## Software

The software that is used for data mining is designed specifically to analyze the relationship within the data and the patterns that emerge from the data according to the needs of the business. For example, data mining software could be used to develop distinct classes of information. Let us take a look at how a grocery store could use data mining. The manager could use the software to understand when it would be the right time to hold a sale for a particular product. The software refers to the data that has been captured and will give the manager a clear idea on the customers' needs and likes. In some instances, data mining specialists may look for clusters of data according to logical relationships, or they study associations and patterns to make conclusions about consumer trends.

## Strategies

There are different kinds of analysis that can be performed by businesses in order to obtain valuable data. There are different types of business data analytics that could be used to obtain different results each of which will impact the business differently. The type of strategy that needs to be used depends fully on the business and the kind of problem that the business is trying to address.

There are different forms of data analytics that could result in different outcomes thereby offering different insights on the data obtained by the business. One of the best ways to retrieve insights from data is through the process of data mining.

When it comes to developing the data analytics strategy, it is important to understand the definition of data mining and also understand how crucial it is to your business. It is important to take note that the goal of any data mining process is to look for relevant information that would be easily understood when looking at large scale data sets.

Below are the most common types of data mining analytics that you can use for your business.

## *Anomaly Detection*

Anomaly detection is the process of searching for information in a data set that does not match the expected behavior of the pattern that was predicted. Anomalies are also known as contaminants, exceptions, surprises and most commonly, outliers. They often offer information that is crucial. Outliers are certain parts of the data set that deviate considerably from the general average of the data set thereby causing a hindrance to the analysis. When looked at in numerical terms, these outliers are separate from the rest of the data and could signify that further analysis would need to be made to better understand the data.

Detecting anomalies in data sets could be used to identify if there are risks or frauds that are taking place within crucial systems. These outliers or anomalies have properties that would attract the attention of a data analyst who would then delve deeper to understand what is happening within the data set. This will help the business identify a number of crucial situations that indicate fraud, flaws in processes and also flaws in strategies used by the business.

It is important to take note that in large data sets, there are definitely going to be some anomalies. It is true that anomalies show bad data, but it is also true that these anomalies could be caused due to random variation and may not affect the data badly. In such situations, it is necessary to conduct more analysis.

## *Association Analysis*

Association analysis helps the business identify the associations that exist between different variables in a large-scale data set. This strategy will allow the data analyst discover patterns that are concealed within the data which would help to identify variables inside the data and also understand the occurrences of some variables within the data set.

This strategy is commonly used in retail stores to identify patterns that are found within the data set or information from POS. These patterns can come in handy when it comes to identifying or recommending new products to other managers of the same retail store or to loyal customers based on what has been purchased before at the store. If you do this correctly, you will be able to increase your business's conversion rate and also increase your profits.

A good example would be that of Walmart's use of data mining since 2004. The retail giant discovered that the sale of Strawberry pops had increased by at least seven times before the hurricane. In response to this finding, the retail store placed this product at checkout counters when a hurricane was to strike in a particular area.

### *Regression Analysis*

Regression analysis is used to determine the dependency, if any, between the different attributes in a data set. There is an assumption that one variable could elicit a response from another variable thereby causing a change in the data set.

Attributes that are often assumed to be independent could be affected by each other, which would mean that a dependency is created between them. By using regression analysis, the business owner will be able to identify if one variable is dependent on another variable and if there is any mutual dependency between the two.

A business can use regression analysis to identify and determine the levels of satisfaction of a client and how well this attribute can impact customer loyalty and also how the service could be affected.

Another good example to look at is the information collected by dating sites. They use regression analysis to offer better services to their users. For instance, if you have a Tinder account, the data provided by you is matched against the data provided by another member on tinder. If there is any dependency between these variables, you will find that person's profile on your feed.

Data mining could help businesses look for and focus on the most relevant and important information, which could be used to establish models that could help in making projections on how systems or people could behave so the business could make some projections.

When more data is gathered, better models can be built which would help you use the strategies in better ways resulting in a greater value for your business.

## *Clustering Analysis*

The process of detecting data sets within data sets that have similar attributes is called clustering analysis. This type of analysis helps you identify the similarities and the differences between the data collected. Every cluster has a specific trait that would be used to enhance any algorithms that have been created by the business. For example, there could be customers who would often purchase the same products from the store and these customers could be grouped into a cluster. This cluster could be targeted and analyzed by the data analysts to increase the revenue for a firm.

An outcome of clustering analysis is the development of customer personas that can then be used to represent different customer types within a particular data set. This includes the behavior set or attitude of customers who are actually using the brands or products. The business can use a particular software or programming language to work on relevant cluster analysis.

# Machine Learning

# Chapter 3: What is Machine Learning?

It is not an easy task to define learning since it includes a wide range of processes. If one were to look into a dictionary for definitions of learning, one would come across phrases such as "to gain knowledge, or understanding of, or skill in, by study, instruction, or experience," and "modification of a behavioral tendency by experience." In a manner similar to the way psychologists study learning in humans, this book will focus on the learning processes of machines. Certainly, the two spheres of animal learning and machine learning are intertwined. A few techniques in machine learning are derived from some techniques used in the former. Also, breakthroughs in machine learning can help in bringing to light a few facets of biological learning.

When it comes to machines, speaking superficially, it can be said that any change to a machine's structure, data stored in memory or composition, in order to make the machine perform better and have better efficiency, can be the sign of learning in a machine. However, when we go deeper into the field, only a few of these changes can be subsumed into the category of machine learning. For instance, consider a machine that is meant to predict weather forecasts in a certain area for a few weeks. If data about the weather in the area over the past year is added to the memory of the machine, the machine can learn from this data and predict the weather more accurately. This instance can be called machine learning.

To be precise, the field of machine learning is applicable to machines that are associated with artificial intelligence. Machines associated with artificial intelligence are responsible for tasks such as prediction, diagnosis, recognition, and others. These types of machines "learn" from data that is added to them. This data is called training data because it is used to train the machine. The machines analyze patterns in this data and use these patterns to perform their actions. Machines use different learning mechanisms to analyze the data depending on the actions they are required to perform. These mechanisms can be classified into 2 broad categories- supervised learning and unsupervised learning.

Skeptics of the field of machine learning might question why machines need to learn in the first place. People may wonder why machines aren't designed specific to the tasks that they need to carry out. There are many reasons as to why machine learning is advantageous. As mentioned earlier, research into the field of machine learning can help us better understand certain aspects of human learning.

Also, machine learning can also increase the accuracy and efficiency of machines. A few other reasons are:

- Even with the greatest efforts by engineers, some tasks cannot be defined explicitly. Some tasks need to be explained to the machine using examples. The idea is to train the machine with the input and teach it how to reach the output. This way, the machine will know how to deal with future inputs and process them to reach appropriate outputs.
- The field of machine learning is also intertwined with the field of data mining. Data mining is essentially the process of looking through loads and loads of data in order to find important correlations and relationships. This is another advantage of machine learning in that it might lead to the finding of important information.

- On many occasions, it is possible that humans design machines without correctly estimating the conditions in which it will be functioning. Surrounding conditions can play a huge role on the performance of the machine. In such cases, machine learning can help in the acclimation the machine to its environment so that the performance is not hindered. It is also possible that environmental changes might occur and machine learning will help the machine to adapt to these changes without losing out on performance.
- Another loophole in the process of human beings hardcoding the process into the machine is that the process might be extremely elaborate. In such a case, the programmer might miss out on a few details since it would be a very tedious job to encode all the details. So, it is much more desirable to allow the machine to learn such processes.

- There are constant changes in technology in the world. Major changes occur in other aspects as well such as vocabulary. Redesigning systems to accommodate for each and every change is not practical. Instead, machine-learning methods can be used in order to train the machines to adapt to these changes.

## Subjects involved in machine learning

The sphere of machine learning is the intersection of many subjects. It derives information from a number of subjects. Each of these subjects brings in new methodology and terminology. All of these concepts come together to form the discipline of machine learning. Given below are a few of the subjects involved in machine learning.

### *Statistics*

One of the most common problems in statistics that is used in machine learning is training, i.e.

the usage of samples taken from an unknown probability distribution in order to predict what distribution a new sample is picked from. Another related problem is the estimation of the value of a function at a certain point based on the value of the function at a few sample points. Solutions to these problems are instances of machine learning because the problems involve estimation of future events based on data about past events. Statistics forms an extremely important part of machine learning.

### *Brain modeling*

The part of brain modeling that is closely related to machine learning is the concept of neural networks. Scientists have suggested that one possible model for neurons or the neural network is non-linear components with weighted inputs. Extensive studies have been conducted on these non-linear elements in recent times. Scientists concerned with brain modeling are interested in gaining information about human learning from the study of these networks of neurons. Connectionism, sub-symbolic processing, and brain style computation are a few spheres that are associated with these types of studies.

### *Adaptive control theory*

Control theory is a subject associated with the control of systems. A major problem faced by systems is the change in environmental conditions. Adaptive control theory is a part of this subject that deals with the methods by which systems can adapt to these changes and continue to perform optimally. The main idea is that the systems should anticipate these changes and modify themselves accordingly.

### *Psychological modeling*

For many years now, psychologists have tried to understand the learning processes of humans. One such example is the EPAM network. This network was used for storing and retrieving one of two words when provided with the other. Later on, the concepts of decision trees and semantic networks were conceived in this field. In recent times, the work in the field of psychology has been strongly influenced by the subject of artificial intelligence. Another aspect of psychology that has been studied in recent times is reinforcement learning. This concept has also been used in machine learning extensively.

### *Artificial intelligence*

As mentioned earlier, a large part of machine learning is concerned with the subject of artificial intelligence. Studies in artificial intelligence have focused on the usage of analogies for learning purposes and also on how past experiences can help in anticipating and accommodating future events. In recent years, studies have focused on devising rules for systems that use the concepts of inductive logic programming and decision tree methods.

***Evolutionary models***

It is a common idea in evolutionary studies that not only do animals learn to perform better in life, but they also learn to better adapt to their surroundings in order to enhance their performance. As far as machines are concerned, the concepts of learning and evolving can be considered to be synonymous with each other. Therefore, models that have been used to explain evolution can be used to devise machine-learning techniques. The most prominent technique that has been developed using evolutionary models is the genetic algorithm.

## Varieties of Machine Learning

So far, this book has given an introduction to machine learning and has answered the question about the subjects that constitute it. Now, we come to the more important question of what can be learned in the subject of machine learning. The following are a few topics on which knowledge can be gained through the study of machine learning:

- Programs and logic rule sets
- Terminology and grammars
- Finite state machines
- Problem - solving systems
- Functions
- Artificial Intelligence
- Statistics

Out of the above, the two most focused on topics are those of statistics and artificial intelligence. These 2 subjects are used extensively in machine learning. We now move on to chapters that describe the two broad categories of machine learning techniques: supervised machine learning and unsupervised machine learning.

## Uses of Machine Learning

Machine Learning is now a solution to complete manual tasks that are impossible to complete over a short span of time for a large amount of data. In this decade, we are overcome with data and information and have no manual way of processing this information paving a way for automated processes and machines to do that job for us.

Useful information can be derived when the process of analysis and discovery becomes automated. This will help us drive our future actions in an automated process. We have therefore come into the world of Big Data, Business Analytics, and Data Science. Predictive analytics and Business Intelligence are no longer just for the elite but also for small companies and businesses. This has given these small businesses a chance to participate in the process of collecting and utilizing information effectively.

Let us now take a look at some technical uses of machine learning and see how these uses can be applied to real world problems.

### *Density Estimation*

This use of machine learning allows the system to use the data that is provided to create a product that looks similar to it. For instance, if you were to pick up the novel War and Peace from the shelves of a bookstore and run it through a machine, you will be able to make the machine determine the density of the words in the book and provide you with work that is exactly similar to War and Peace.

### *Latent Variables*

When you work with latent variables, the machine uses the method of clustering to determine whether the variables are related to one another. This is a useful tool when you do not know what the cause of change in different variables is and also when you do not know the relationship between variables. Additionally, when the data set is large, it is better to look for latent variables since that helps to comprehend the data obtained.

### *Reduction of Dimensionality*

Most often, data obtained has a number of variables and dimensions. If there are more than three dimensions, it is impossible for the human mind to visualize the data. It is in these instances that machine learning can help in reducing the data into a manageable number of dimensions so that the user understands the relationship between the variables easily.

### *Visualization*

There are times when the user would just like to visualize the relationship that exists between variables or would like to obtain the summary of the data in a visual form. Machine learning assists in both these processes by summarizing the data for the user using specified or non-specified parameters.

## Applications of Machine Learning in the Real World

There are three areas of knowledge where machine learning can be used effectively – business, web mining and scientific applications.

### *Business Applications*

Machine learning can be used to profile customers, which would help businesses make better products that would cater to each segment of the customer base. Banks and other financial companies can detect frauds and can track the spending and saving patterns of customers to identify if there is something wrong with the account. For instance, in the TV

Series 'Friends,' the credit card company called to confirm with Rachel if she was alright since she had not been using her card for some time. This will help companies retain their customers since they will begin to understand their customers better.

Businesses can also use machine learning to control and analyze the processes being used. For instance, businesses can track the amount of time spent by each sales executive over the phone and identify the sales that they have been able to make. Most businesses can use machine learning to understand their customers and also help to provide their customers with the right recommendations. There are a number of different ways that unsupervised and supervised machine learning can be used by businesses, but these are often the most basic uses.

## *Web Mining*

Machine learning is great for document searching and text mining. You can look for a particular object or organize all the information you have with respect to a particular object. Another example of supervised and unsupervised machine learning would be social media. Every social media platform keeps a track of the user's browsing and would provide the user with information or links that he or she would prefer to look at. The best example would be the recent update provided by Instagram where the platform would show the users the pictures that they would love the most at the top of their feed.

## *Scientific Applications*

Machine learning can be used in a number of different applications right from genetics to pharmacology. In pharmacology, machine learning is used to identify and discover newer medicine and techniques to screen and prepare drugs. In medicine it is used to scan ECGs, MRIs and any other tests to identify any irregularities in the patterns obtained as a result of those tests. Machine learning can be used in satellite and probe data to understand the geological data that is obtained. Predictions can be made about any natural calamities that may occur using these data.

These applications of machine learning only graze the surface of the world of capabilities of machines to gather information. There are a number of other applications of machine learning which we are uncovering every single day.

# Chapter 4: Supervised Machine Learning

As mentioned earlier, an important process of machine learning is called training where the machine is fed with data about past events so that the machine can anticipate future events. When this training data is supervised, it is called supervised machine learning. The data fed essentially consists of training examples. These examples consist of inputs and the desired outputs. These desired outputs are also known as supervisory signals. The machine uses a supervised learning algorithm, which generates an inferred function used to forecast events. If the outputs are discrete, the function is called a classifier and if the outputs are continuous, the function is known as a regression function. This function is responsible for predicting outputs of future inputs. The algorithm needs to conceive a generalized method of reaching the output from the input based on the previous data. An analogy that can be made in the spheres of human and animal learning is concept learning.

## Overview

Supervised learning is a method that uses a fixed algorithm. Given below are the steps involved in this algorithm:

1. The first and foremost step in supervised learning is the determination of the type of examples to be used for training the machine. This is an extremely crucial step and the engineer needs to be very careful in deciding the kind of data he wants to use as examples. For instance, for a speech recognition system, the engineer could either use single words, small sentences or entire paragraphs for training the machine.

2. Once the engineer has decided on the type of data he wants to use, he needs to collect data in order to form a training set. This set needs to be representative of all the possibilities of that function. So, the second step requires the engineer to collect inputs and desired outputs for the training process.
3. Now, the next step is to determine how to represent the input data to the machine. This is very important since the accuracy of the machine depends on the input representation of the function. Normally, the representation is done in the form of a vector. This vector generally contains information about various characteristic features of the input. However, the vector should not include information on too many features since this would increase the time taken for training. A larger number of features might also lead to mistakes made by the machine in prediction. The

vector needs to contain exactly enough data to predict outputs.

4. After deciding on the representation of input data, a decision must be made on the structure of the function. The learning algorithm to be used must also be decided on. Most commonly used algorithms are decision trees or support vector machines.

5. Now the engineer must complete the design. The learning algorithm chosen should be run on the data set that has been gathered for training. Sometimes, certain algorithms require the engineer to decide on some control parameters in order to ensure that the algorithm works well. These parameters can be estimated by testing on a smaller subset or by using the method of cross validation.

6. After running the algorithm and generating the function, the accuracy of the function should be calculated. For this, engineers generally use a testing

set. This set of data is different from the training data and the corresponding outputs to the inputs are already known. The test set inputs are sent into the machine and the outputs obtained are checked with those in the test set.

There are a number of supervised learning algorithms in use and each one has its own strengths and weaknesses. Since there is no definitive algorithm that can be used for all instances, the selection of the learning algorithm is a major step in the procedure.

## Issues to consider in Supervised Learning:

With the usage of supervised learning algorithms, there arise a few issues associated with it. Given below are 4 major issues:

### *Bias-variance tradeoff*

The first issue that needs to be kept in mind while working with machine learning is the bias-variance tradeoff. Consider a situation where we have various different but equally good training sets. If, when a machine is trained with different data sets, it gives systematically incorrect output predictions for a certain output, the learning algorithm is said to be biased towards that input. A learning algorithm can also be considered to have a high variance for an input. This occurs when the algorithm causes the machine to predict different outputs for that input in each training set. The sum of the bias and variance of the learning algorithm is known as the prediction error for the classifier function. There exists an exchange between variance and bias. A requirement for learning algorithms with low bias is that they need to be flexible enough in order to accommodate all the data sets. However, if they are too flexible, the learning algorithms might end up giving varying outputs for each training set and hence increases the variance. This kind of learning methods needs to be able to amend according to this tradeoff. This is generally done spontaneously or by using an adjustable parameter.

## *Function complexity and amount of training data*

The second issue is concerned with deciding on the amount of training data based on the complexity of the classifier or regression function to be generated. Suppose the function to be generated is simple, a learning algorithm that is relatively inflexible with low variance and high bias will be able to learn from a small amount of training data. However, on many occasions, the function will be complex. This can be the case due to a large number of input features being involved or due to the machine being expected to behave differently for different parts of the input vector. In such cases, the function can only be learned from a large amount of training data. These cases also require the algorithms used to be flexible with low bias and high variance. Therefore, efficient learning algorithms automatically arrive at a tradeoff for the bias and variance depending on the complexity of the function and the amount of training data required.

## *Dimensionality of the input space*

Yet another issue that needs to be dealt with is the dimensionality of the input vector space. If the input vector includes a large number of features, the learning problem will become difficult even if the function only considers a few of these features as valuable inputs. This is simply because the extra and unnecessary dimensions could lead to confusion and could cause the learning algorithm to have high variance. So, when the input dimensions are large, the classifier is generally adjusted to offset the effects by having low variance and high bias. In practice, the engineer could manually remove the irrelevant features in order to improve the accuracy and efficiency of the learning algorithm. However, this might not always be a practical solution. In recent times, many algorithms have been developed which are capable of removing unnecessary features and retaining only the relevant ones. This concept is known as dimensionality reduction, which helps in mapping input data into lower dimensions in order to improve the performance of the learning algorithm.

### *Noise in the output values*

The final issue on this list is concerned with the interference of noise in the desired output values. It is possible that the values of the desired outputs (supervisory targets) can be wrong due to the noise that gets added at sensors. These values could also be wrong due to human error. In such cases, the learning algorithm should not look to match the training inputs with their exact outputs. For such cases, algorithms with high bias and low variance are desirable.

### *Other factors to consider*

- One important thing to be kept in mind is the heterogeneity of data. The level of heterogeneity of the data should also play a role in dictating the learning algorithm that is to be chosen. Some algorithms work better on data sets whose inputs are limited within small ranges. A few of these are support vector machines, logistic regression, neural networks, linear regression and nearest neighbor methods. Nearest neighbor methods and support vector machines with Gaussian kernels work especially better with inputs limited to small ranges. On the other hand, there exist algorithms like decision trees, which work very well with heterogeneous data sets.

- Another feature of the data sets that needs to be considered is the amount of redundancy in the set. A few algorithms perform poorly in the presence of excessive redundancy. This happens due to numerical instabilities. Examples of these types of algorithms are logistic regression, linear regression, and distance based methods. For such cases, regularization needs to be included so that the algorithms can perform better.

- While choosing algorithms, engineers need to consider the amount of non-linearities in the inputs and the interactions within different features of the input vector. If there is little to no interaction and each feature contributes independently to the output, algorithms based on distance functions and linear functions perform very efficiently. However, when there are a number of interactions within the input features, algorithms based on decision trees and neural networks are desirable. The reason for this is that these algorithms are designed to detect these interactions in the input vectors. If the engineer decides to use linear algorithms, he must specify the interactions that exist.

When an engineer is tasked with selecting an algorithm for a specific application, he may choose to compare various algorithms experimentally in order to decide which one is best suited for the application. However, a large amount of time needs to be invested by the engineer in collecting training data and tuning the algorithm. If provided with a large number of resources, it is advisable to spend more time collecting data than spending time on tuning the algorithm because the latter is extremely tedious. The most commonly used learning algorithms are neural networks, nearest neighbor algorithms, linear and logistic regressions, support vector machines and decision trees.

# Chapter 5: Unsupervised Machine Learning

At this point, the reader should be familiar with the concept of supervised machine learning wherein the machine is trained using sets of inputs and outputs that are desired. However, there are other techniques of machine learning. One of these is known as reinforcement learning. In this technique, the machine is designed to interact with its ambient environment through actions. Based on the environment's response to these actions, the machine receives rewards if the environment reacts positively or punishments if it reacts negatively. The machine learns from these reactions and is taught to perform in a manner such that it can maximize the rewards it will obtain in the future. The objective could also be to minimize future punishments. This technique of learning is related to the subjects of control theory in engineering and decision theory in statistics and management sciences.

The main problems studied in these two subjects are more or less equivalent and the solutions are similar as well. However, the two subjects focus on different parts of the problem. There exists another technique of machine learning that is closely related to game theory and also uses reinforcement learning. The idea here is similar to that in reinforcement learning. The machine produces some actions that affect the surrounding environment and it receives rewards or punishments depending on the reaction of the environment. However, the main difference is that the environment is not static. It is dynamic and can include other machines as well. These other machines are also capable of producing actions and receiving rewards (or punishments). So, the objective of the machine is to maximize its future rewards (or minimize its future punishments) taking into account the effects of the other machines in the surroundings.

The application of game theory to such a situation with multiple, dynamic systems is a popular area of research. Finally, the fourth technique is called unsupervised machine learning. In this technique, the machine receives training inputs but it does not receive any target outputs or rewards and punishments for its actions. This begs the question- how can the machine possibly learn anything without receiving any feedback from the environment or having information about target outputs? However, the idea is to develop a structure in the machine to build representations of the input vectors in such a manner that they can be used for other applications such as prediction and decision-making. Essentially, unsupervised learning can be looked at as the machine identifying patterns in input data that would normally go unnoticed. Two of the most popular examples of unsupervised learning are dimensionality reduction and clustering. The technique of unsupervised learning is closely related to the fields of information theory and statistics.

# Artificial Intelligence and Data Science

## Chapter 6: Artificial Intelligence 101

An amazing development has taken place over the last few years. You may have watched how robots in Star Wars were able to perform so many actions and maybe, just maybe, you would have wondered how wonderful it would have been if this were to happen in this decade. You may not have seen it coming, but this was an inevitable turn of events – the emergence of Artificial Intelligence (AI). Everywhere we look today, we come across a number of intelligent systems that talk to us – Siri, Google Assistant – offering us advice and also offering us recommendations. These systems improve almost every year to improve its interpretation of images, voice recognitions and also to drive cars based on different techniques used by Facebook and Google's Deep Learning Efforts. Other work always aims to understand and generate machines that would understand our language and communicate with us.

The reemergence of AI has caused a lot of confusion since there are so many companies that have begun to explore the scene. How do we make sense of any of it?

Let us start with a simple definition of AI. Artificial Intelligence or AI is a field of computer science that is aimed at developing computers that are capable of performing tasks that can be done by people, especially those tasks that are considered to be performed by intelligent people.

AI has had a number of excellent runs. In the early sixties, there were great promises that were made about what a machine could do. In the eighties, it was said to revolutionize the way businesses were run. But in those eras, the promises that were made were too difficult to deliver. So what makes the latest developments in AI any different? What makes the systems developed now any different from the diagnostic programs and neural nets of the past? There are a number of reasons why the developments in this era are different from the last.

## Increased Computational Resources

The computers we have in this era are faster and can think harder thereby increasing the computational power. The techniques used earlier worked well only in the past, but now there would be a necessity to improve the computational grid and also expand it.

## Growth of Data

Over the years, the data collected has increased by a vast amount and this data is being made available to the machines. This increase in the data has given machines much more to think about. This goes to say that learning systems get better at understanding more data and would now be able to look at thousand examples as opposed to only a few hundred.

## Deeper Focus

AI has shifted away from looking at smaller aspects of data to look at specific problems. The systems now are capable of thinking about a particular problem as opposed to daydreaming without any problem. Systems like Cortana and Siri work well within limited domains that can be focused and modeled on pulling specific words that you would have said instead of understanding the entire sentence spoken by you.

## Knowledge Engineering

The problem and issues with knowledge engineering have been transformed into different aspects of learning. The systems developed these days use their own ways to learn. The bottleneck in the systems in the past helped us add more rules to avoid such bottlenecks in the further processing of data. Most approaches in the modern times focus on learning these rules automatically.

## Alternative Reasoning Models

Alternative reasoning models have been adopted which are based on the assumption that the systems do not have to reason like human beings in order to be smart. The machine is allowed to think like a machine.

It is these factors put together that have given the world the first renaissance of intelligent machines that have become a part of our lives and have been adopted in different tools and in workplaces.

## Exploring AI

The goal for some developers and researchers is to build a system that has the ability to act and think intelligently the way human beings do. Most others do not care if the system does think like human beings – they are only worried about whether or not the system would do the right thing. Apart from these two schools of thought, there are others that are somewhere in between. If we were to use human reasoning in systems, it would help computers do things similar to human beings.

### *Strong AI*

The work that is aimed at stimulating human reasoning in a machine is what is called strong AI. This implies that any machine created would need to have the capabilities of obtaining the results and also explain how a human being would think. Genuine models of AI systems that are simulations of the human mind and cognition have to be built.

### *Weak AI*

The work that was done within the second school of thought, aimed towards just getting structures to work, is commonly known as weak AI in that even as we might be able to build structures which can behave like people, the outcomes tell us nothing about how human beings think or feel. One of the best examples of this is IBM's Deep Blue, a gadget that became a master chess player but absolutely did not play in the same manner that human beings do and informed us very little about cognition in general.

### *Anything in Between*

The systems in between Weak and Strong AI are those that are informed and understand human reasoning but are not slaves to it. This is where most of the powerful work is happening in AI today. This work uses human emotion and reasoning as a guide but it does not use it as a goal to model the system. If only one were able to come up with a name for this school of thought – maybe Practical AI.

The important takeaway is that a system does not have to be smart the way human beings are smart in order for it to be AI. The system would just need to be smart.

# Chapter 7: Big Data and Artificial Intelligence

Most AI Systems have started to use three components of human reasoning – assessing, inferring and predicting. There are a number of questions that would need to be answered every day – what is happening around us, what it means, what will happen next and so on. When we push the call button, walk towards and elevator or wait for a taxi, we are trying to assess, infer and predict. AI systems have started to do the same thing, although they would not wait in line to catch a taxi.

In the world of Big Data, where at least 2.5 quintillion bytes of data are collected or produced every single day, it important to know how AI Systems would capture that data, synthesize it and use it to drive their reasoning.

The use of AI systems in issues like Big Data allows us to transform the world of numbers and unstructured data into knowledge that can be used by human beings.

The trick to understanding these systems is to understand that the processes that underlie the intelligence that are smart. The systems are built on a foundation that processes are not only simple but also understandable. It is important to remember that AI is not magic but it is the application of algorithms that are powered by data, processing power and scale.

## What is Big Data?

Data that is obtained from two significantly different sources is grouped under the banner of Big Data. The first source of data is the data that is found within the organization or business and that is shared across a vast network. This data includes PDF documents, blogs, emails, work documents, business events, internal files, process events and any other structured data, unstructured or semi-structured data that can be found within the organization. The second source of data would be the data that is available outside the organization. Most of this information is available for free but there is some information that would require a certain payment to be made in order to obtain it. Most of this information is available to the general public, but there is some information that is considered confidential and is kept within the organization itself. This information includes literature on products that is distributed by the competitors, information available on social media, hints from third parties, certain organization hierarchies and also any complaints that may have been posted by customers on regulatory sites.

You may begin to wonder what makes Big Data any different from the data that is being used since the ancient times. There are four characteristics of Big Data that make is very different from other kinds of data – Volume, Velocity, Variety, and Veracity. There are a number of other characteristics that differentiate Big Data from other types of data, but these are the most important and prominent characteristics to consider.

## *Volume*

Numerous organizations have started to struggle with the amount of data that is being stored in their data warehouses since big data took the world by a storm. It was noted in the Fortune magazine that the world had already created close to 10 exabytes of data by the year 2006 and that this number would double by the year 2015. It has also been identified that this amount of data can now be collected within a few nanoseconds thereby increasing the data collected by trillion gigabytes every hour, which is a scary amount of data.

A few years ago, there were organizations that would count their data storage space for any type of analytics in terabytes. Now, they can only do the same in petabytes since the data collected has doubled or tripled since then. The amount of data does cause a strain on the analytics architecture for different organizations in different industries. For instance, let us consider an organization that is in the communication industry. You know that this organization would definitely have a hundred million customers. If data were collected for these customers on a daily basis, there would be at least 5 petabytes of data in a hundred days. Most companies try to get rid of data daily, but regulators have asked most companies in the communication industry to store records about calls made or the data used by each customer.

## *Velocity*

Velocity can be understood from two different perspectives – one that represents latency while the other represents the throughput of data. Let us first take a look at latency. The analytics infrastructure was once a store – and – report environment where the data that was collected the previous day was used to make reports and this data was represented as "D-1." Over the last decade, the analytics infrastructure has been used in different business processes that there was a necessity to change the infrastructure to cater to the needs of all businesses. For example, some advertising agencies are trying to conduct analytics to help them place advertisements on online platforms within 10 milliseconds.

Let us now consider the second measure of velocity – the throughput of data. This represents the data that is flowing through the pipes of the analytics infrastructure. The amount of global mobile data is growing at the rate of 80 percent, which is said to compound annually implying that the data that is collected annually is set to increase by 12 – 14 exabytes per year as users begin to share more images and videos with the world. In order to analyze this data, many corporates are seeking analytical infrastructure that will help them process information in parallel.

### *Variety*

It was in the late 1990s that Data Warehouse Technology was introduced to create and represent data using Meta – models that would help to simplify data and also help to represent data collected in one form, irrespective of whether or not the data was structured.

The data was often compiled from multiple sources and was then transformed using either ETL (Extract, Transform, Load) or ELT (Extract data, load it into the warehouse and transform the data within the warehouse). The idea was to narrow down the variety and ambiguity in the data collected and also structure that data to put it to good use. Big Data has helped to expand our horizons by enabling new data integration and solutions infrastructure and data analytics technology. A number of call center analytics constantly seek solutions that would help them attend to their customers better. These solutions would also help them understand the conversation that takes place between them and the disgruntled customer, which would then give them an idea of how to proceed further. The source data that is obtained from call centers are only unstructured data like audio, sound, or text and some structured data. Different applications gather different types of data from blogs, emails, and documents. For example,

Slice, a company used to provide analytics for online orders, uses raw data that comes from a number of different organizations like online bookstore purchases, airline tickets, parking tickets, music download receipts and any other purchases that may have hit your email. How can this information be normalized into creating product catalogs and also analyze any purchases made?

### *Veracity*

Internal data is often carefully governed while Big Data comes from a number of sources that are outside the control of the company and this data often comes with a lot of inaccuracies and incorrectness, which would hamper the analysis made. Veracity represents the sustainability of the data obtained and also looks for the credibility of the sources of data that is obtained for a target audience.

Let us try to see how we can understand the credibility of the sources of data. If organizations began to collect information on their own products using third parties and then provide that information to their customer care support or their contact center employees in order to support the queries that customers have, the organizations providing the information would need to be assessed and screened before trusting the information that is provided by them. Otherwise, there can be a possibility of the organization making a wrong offer to the customer thereby affecting the revenue. Numerous social media responses to different campaigns could come from unhappy customers or disgruntled employees, which would definitely impact the decisions made by the company. For instance, if there was a survey conducted by companies on the products provided by them and customers happened to like the product, this would be taken into consideration. However, if the customer were to select an option that did not show that they liked the product, it is important to know why before taking this unstructured piece of data into account.

We have to always consider the amount of truth that can be shared with an audience. The veracity of the data that is collected or created within a given organization is often considered to be well intentioned. However, some of that data cannot be shared with the public since there could be some threat to the company. This would reduce the chance of having a wider communication network. For instance, if customer service has provided the manufacturing and engineering department with the shortcomings of a particular product based on the feedback given by customers, the data shared should be selective and on a need – to – know basis only. Other data can be shared with some customers who have certain prerequisites or contracts.

Big Data makes it possible for numerous businesses to process all the information that is collected on the simple learning systems. They would need to identify one signal from all the noise. It is only through this way of learning that it is possible to translate sentences from one language to the other. Parallelism and processing enable systems to take up thousands of pieces of evidence and then test them independently before adding up the results.

## Assessing Data using AI

Consumer systems always try to assess human beings. For examples, Amazon puts together a picture of who you are and what products interest you and also assesses you against other similar customers and makes predictions about you. The data that is used to make this prediction is transactional – what you look at and what you purchase. Amazon's recommendation engines use all this information to come to a conclusion of what products should be recommended to you.

Profile data is only a small part of the picture. To this data, information about different categories – cluster objects (DIY books), Customers and other information – would need to be assessed to make an informed decision. The area you live in and the amount of money you would be willing to spend can be pulled in to refine the assumptions made about you.

The results are always a set of characterizations:
- Based on what collections you like, you have a lot in common with a person across the globe that has a similar interest in science fiction, DIY books, and cookbooks and so on.
- Based on a set of things you were looking at, you could be someone who loves to garden.
- You just bought gardening tools to create your own kitchen garden.

For retailers, the information that is of utmost importance is transactions, product categories and the clusters of people. For search engines, the important information is the history of the information you have looked for and also the items you have clicked on and at which location those clicks were made.

# Chapter 8: How does AI draw conclusions from Data?

There are different methods that can be used by learning systems to infer something from data. This chapter covers the most common methods used for the same.

## Deductive Reasoning

Deductive reasoning ensures that the system comes up with a distinct solution. The process starts with the assertion of a rule or assumption and then proceeds from there to arrive at the conclusion. The assumption made in this form of reasoning is that if the original rule or assumption made is true, then the conclusion is definitely true. For instance,

If $x = 3$

And $y = 4$

Then, $4x+2y = 20$

In the above example, there exists a logical necessity that (4x+2y) is equal to 20. The symbolic logic in the above equations uses a language that helps the system come to this conclusion. But, if there were no equations and only English sentences were being used, there would be no difficulty to come to a conclusion if the system were to use deductive syllogism. For instance,

*The disorder in a system would increase unless energy is used. If my bedroom were a system, then disorder would constantly increase unless I was to clean the room.*

In the syllogism given above, the first statement and the second statement lead to the conclusion that the room would need to be cleaned to avoid having any disorder in it. Consider another example:

*Medical technology, if used successfully to treat patients, should be funded. Stem cells have been used to treat patients successfully using sixty new therapies. Stem cell research and technology should be funded.*

A conclusion that is made is considered true or false depending on the premise of the argument. However, the deductive inference can be considered true or false irrespective of whether or not the premise is true. The final conclusion can be valid even if the premise is false. For instance,

*Drought does not exist in the West.*
*Las Vegas is in the West.*
*People living in Las Vegas do not have to worry about dealing with drought.*

The inferential process is valid and the conclusion obtained based on the premises is also valid. However, this conclusion is unsound since the premise itself is unsound. This form of a syllogism is very deceiving since it looks very logical and to be very fair, it is logical. But, if the premise were incorrect, the conclusion obtained would never serve the purpose of the exercise.

If the assumption were made that propositions are sound, the stern logic and process followed in deductive reasoning would give conclusions that are absolutely certain. Deductive reasoning cannot enhance the workings of an AI system because the conclusions are often self – evident. The system would be unable to make a prediction about the future.

## Inductive Reasoning

Inductive reasoning starts with making observations that are limited to the scope of the issue and ends with making a generalized conclusion, one that is likely but is not certain since the conclusion is based on accumulated evidence. Inductive reasoning moves from specific conclusions to general conclusions. AI systems often use inductive reasoning to obtain conclusions. They gather evidence, seek patterns and form hypotheses or theories to explain what has occurred.

Conclusions that are obtained using inductive methods of reasoning need not be logical. No amount of evidence obtained would guarantee the conclusion. This is for the reason that there is absolutely no way to confirm that all the evidence necessary to derive the conclusion has been drawn and that there is no other unobserved evidence that exists to validate that hypothesis. Newspapers could report a number of conclusions that could have been obtained by AI but there is an important aspect to consider – AI can only envision what could possibly happen, it cannot be certain of the outcome of events when using inductive reasoning.

Since inductive conclusions need not be logical, the arguments are most often not true. They are cogent, implying that the evidence seems relevant, complete and convincing to an extent that the conclusion could be true. It can also be said that inductive arguments are not false; the right term would be that they are not cogent.

This is an important difference between inductive and deductive reasoning. While inductive reasoning cannot yield absolute conclusions, it can increase the knowledge that the AI system has. It can help the system make predictions about the future.

For instance, Albert Einstein had observed the way a pocket compass moved when he was five years old. He then became fascinated by the idea of something invisible causing the compass needle to move. This observation followed by a number of other observations and the use of mathematical and logical tools helped him derive a conclusion that helped him predict events that were yet to be observed. AI learning system using inductive reasoning can perform the same actions.

## Abductive Reasoning

This type of reasoning begins with a set of observations that are incomplete and then proceeds to form the most likely explanation for that very set of observations. Abductive reasoning helps to derive conclusions that would help to make the right decisions on a daily basis using the information that is at hand, which is often incomplete.

For instance, a doctor would give his diagnosis based on abductive reasoning. He would take a look at the symptoms and try to identify the diagnosis that would explain those very symptoms. The same goes for when jurors listen to a criminal case. The jurors would need to consider the evidence produced by both the defense and the prosecution and make an informed decision covering all points of the evidence collected. There could be no certainty about the verdict since most evidence would not have been admitted to them during the case, but they would make their best guess. The same goes for AI learning systems. They would make informed decisions and come to the right conclusions using whatever information they are provided with.

Inductive reasoning works on the principle that the evidence obtained does not necessarily have to be complete to make a decision. However, when it comes to abductive reasoning, it is important that the evidence obtained is complete in order to make informed decisions or even predict the future. For instance, a doctor may look at only a certain number of symptoms to make his prognosis, but he would be making the best use of those symptoms.

The process of abductive reasoning is creative, intuitive and can also be revolutionary. For instance, Einstein's work was not only deductive and inductive; it also involved a lot of creativity and imagination, which was scarcely warranted at that time. Einstein only made thought-based judgments and it was for this reason that most of his peers did not trust his theories. His theories have proved to be correct to this day!

## Case Based Reasoning

The earlier methods use different premises and assumptions to identify and represent the data collected and then use that data to predict the future. However, in case based reasoning, the premises and assumptions are stored and are accessed to solve future problems. If the system were looking at a new problem, it would look for premises and assumptions that were similar to the new problem and use those cases to predict the target future value. This is one of the easier ways of obtaining a conclusion using an AI system since there is relatively less work that will need to be done offline. The cases used are already stored in the warehouse that can be used by the system.

Regression and classification use case based reasoning. This reasoning is also applicable when the premises and cases are complicated, such as in criminal cases where the cases are often complicated laws and legal rulings and also in planning where the cases are prior solutions to complex problems.

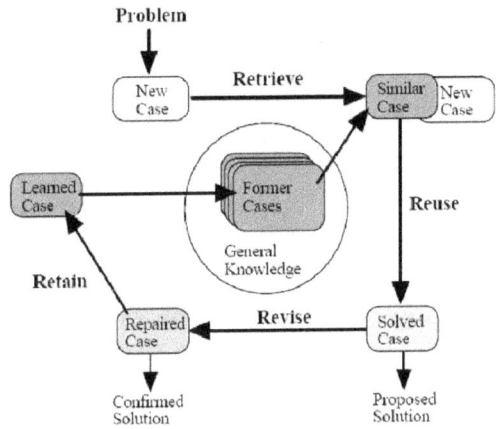

If the cases and premises are simple, only one algorithm can be run to use the case or premise that is similar to the new problem at hand. The system would then create a set of these similar cases and use those as input to predict the target future value.

## Common Sense Reasoning

The ability of human beings to use common sense knowledge is completely dependent on their ability to perform common sense reasoning.

Most artificial intelligence systems have not been designed to infer using the rules of any of the well-known concepts of mathematical logic. There is no distinction between the inferences that are correct and the inferences that would need to be used to solve the problem at hand. However, the logical system often corresponds to the subset of first order logic. Systems often provide for the path to infer a fact about a particular object using other facts and objects that contain variables with general rules. However, many systems never infer statements that are general – they never look for quantified formulae.

Human reasoning involves the process of obtaining facts by making observations in the world and now, computer programs are also capable of doing the same. Robert Filman conducted an interesting thesis on observation in the chess world where he concluded that people obtained many facts using observation although they could use deduction to obtain those very same results. Some smart systems would not need to have the skill to observe, but if we were to build a system that would try to diagnose and draw conclusions from the patient's appearance, it is important to build systems with better observation skills.

An important development in AI is the use of non-monotonic reasoning to make the right judgments and conclusions.

Deductive reasoning is based on mathematical logic and has the property of monotonicity where the system would obtain an analogy using similar mathematical concepts. Let us assume we have a set of premises or assumptions to solve one problem. Now, we add additional premises and assumptions to the set. There could be new conclusions obtained but every sentence that was obtained after a thorough deduction would still be connected to the enlarged set.

Human reasoning does not ordinarily share the monotonicity property. If you knew that I owned a car, you would ask me to give you a ride when you needed one. If you knew that my car was being serviced, which does not change the fact that you knew that I owned a car, you would not ask me for a ride. If I told you that my car would be out in half an hour, you will probably change your mind and ask me again.

A number of artificial intelligence researchers have pointed out that computer programs that are intelligent would need to reason non-monotonically. Based on this observation, some concluded that logic could not be used always.

However, it was seen that deductive reasoning could be substituted with additional modes of non-monotonic reasoning, which are as accurate and precise as deduction and can be studied mathematically. Non-monotonic reasoning has become more formalized and gives rules of conjecture. However these rules of inference imply that the conclusions drawn could be appropriate but could be proved wrong if more facts are obtained.

# Chapter 9: What is Data Science?

Data science is one of the emerging fields in statistics that analogous to the collection, preparation, analysis, visualization, management and preparation of vast amount of information or data. The term is connected to and refers to a number of areas of study including computer science and databases, but there are a set of skills that one would need to possess, including non – mathematical skills, to work as a successful data scientist.

For some, the term data science conjures an image of a statistician in a white lab coat staring at the computer screen and scrolling through numbers. This is far from the truth. Statisticians do not wear coats – you may find them in thick glasses – but this statement of fashion is reserved only for environmentalists, doctors, biologists and scientists who work in environments that are not clothes friendly. Also, the data that is collected is never structured. This implies that the data that is received could be of different kinds – audio, video or even text. Think about a website that is filled with pictures and messages from friends or other users – there is absolutely no way you can find numbers on that page.

It is a known fact that most schools, governments, and companies use enormous amounts of information in the form of numbers – grades, tax, revenues – but there is enormous data that is collected by these organizations that are not structured. The latter form of information makes most statisticians and mathematicians cringe. It is always useful to have people with great mathematical skills, but there is a lot more to data than just math. It is essential to know if people working on data are comfortable with trying to analyze images, words, paragraphs, sounds, videos, lists and various other forms of information.

In addition to the different kinds of data, data science is a lot more than just the analysis and management of information or data collected. There are a number of people who would love to analyze the data collected and spend hours or years looking at plots or histograms, but for those people who prefer to look at more than just plots data science offers a number of roles that would cater to different skill sets. Let us take a look at the different types of data that would need to be considered when it comes to buying a floor cleaning detergent.

Irrespective of what your preference is to buy a detergent – fresh smelling, colorful, good reviews – you prepare yourself to buy this particular commodity when you write it down on your grocery list. You would consider this list to be a small scribble that you will throw out when you finish shopping. When you enter the store, you would remember to buy floor-cleaning detergent because of the list in your hands. At the checkout, the cashier would scan the barcode of the product and the register would log the price. In the data warehouse, the computer would make note that there is one less item available for the same product. This will give the stock manager an idea of how much of the product he would need to buy in the future. If you have a coupon on the product, you would get a discount and the computer will also note this. At the end of the week, the information with respect to the detergent would be uploaded and sent to the detergent company. The company would then pay the discounted amount to the grocery store. At the end of the month, the manager of the store would then create graphs and pie charts, which would show the different detergents sold at the store and how many orders should be placed for the different kinds of detergent.

This tiny piece of data that was on a small sheet of paper has now ended up on different servers in different warehouses. This piece of information would have ended up on the manager's desk and he would need to make a decision on whether or not he would need to place an order for the detergent. This piece of information went from the tiny piece of paper on to the manager's desk. This information would not have been sent in its raw form to the manager's desk – it would have been transformed. In addition to the different computers used and the software used on those computers, other hardware like the barcode scanner comes into picture. All of these put together will help to scan, collect, manipulate, transmit and store the data, which would then be analyzed by the manager.

Data scientists definitely do not take part in each step of the process. They are not a part of the process where hardware would be built to collect the necessary information. You may be wondering where they would be coming into picture. Data scientists play the most active role in these aspects called the four A's of data: Data Architecture, Data Acquisition, Data Analysis and Data Archiving. Let us take a look at the detergent example. With respect to the data architecture, it is important that in the design of the system for sale to think in advance about how the data coming out of the system would be used. The system architect would have a keen appreciation that both the stock and store manager would need to use the data that is scanned at the cash register, for similar purposes. A data scientist would help the architect by providing input on how the data would be used and how it would need to be routed and organized to support the visualization, analysis and presentation of the data to the right people.

Next acquisition focuses on how the data are to be collected and how they would need to be represented before the analysis of the data begins. For instance, every barcode represents a number that does not describe the product itself. At what point must the barcode be associated with the description of the product that it was printed on? Different barcodes are used for the exact same product. When should we make note of the fact that both Products X and Y are the same but have different barcodes? Representation, transforming, grouping and linking of data are all tasks that will need to be performed before the data can be analyzed for profitability. These are the tasks where the data scientist is involved.

The analysis phase is where the data scientist is most involved. In this context, we would be using analysis to summarize the data that was collected, using certain segments of the data to make inferences about the data as a whole. This analysis would then be presented in tables, graphs, charts, and animations. There are many technical, statistical and mathematical aspects to these activities but it is important to remember that the ultimate audience for data analysis is in fact a person or a group of people. These people are the data users and it is these people's needs that data scientists would be fulfilling. This point highlights the need for excellent communication skills in data science. The most sophisticated statistical analysis developed would not be worth anything unless the results that have been created can be communicated effectively to the data user.

Data scientists would need to become actively involved in the process of archiving the data collected and analyzed. The data or information would need to be preserved in a manner that would make the data useful in the future. This could be termed as data curation and the process is a difficult challenge since it is always difficult to anticipate the use of that data in the future. For instance, when developers of twitter were working on how to store tweets, they never anticipated that these tweets could be used to pinpoint earthquakes and tsunamis, but they had enough foresight to realize that these "geocodes" could be useful elements to store as data.

# Python

# Chapter 10: Markov's Model in Python

Now that we have taken a look at Markov Models, Hidden Markov Models, Machine Learning, Artificial Intelligence and Data Science, let us take a look at how to apply our knowledge of the Markov Models in Python.

Python is a simple but powerful language that was developed in the 1980s by Guido van Rossum. This language is a favorite for those who love to code for the purpose of machine learning. Let us first take a look at how to start working in Python before we look at programming Markov Models.

## Getting Started

There are different ways to start working in Python depending on the type of OS you use – Windows, Linus or Mac OSX. This section takes you through the steps you would need to follow for the different systems.

### *Linux*

Step 1: Connect to the Internet.
Step 2: Launch the Terminal Application.
Step 3: Type "su" in the command prompt window and press enter.
Step 4: Type the root password and press enter.
Step 5: There are different ways to install Python on the two types of Linus -

    Debian Based – Type "apt-get install Python" and press enter.

    Redhat/RHEL/Cent OS – Type, "yum install Python" and press enter.

Step 6: Once the installation process is complete, update the system by typing "su-apt get update" and press enter.

***Windows***

Step 1: Go to the website: https://www.python.org/downloads/windows to download Python.

Step 2: Choose the bits based on your Windows software.

Step 3: Open the Python.exe file.

Step 4: Click "Accept Default Settings".

Step 5: Wait for the installation process to complete.

***Mac OSX***

Python 2.7 is already installed on Mac OSX, but if you would like to use the latest version, follow the steps below.

Step 1: Open any of the OSX emulators.

Step 2: Type "/user/bin/ruby_e"$(curl_fsSL http://raw.githubusercontent.com/homebrew/install/master/install)"

Step 3: Wait for the installation process to complete.

Step 4: Go to "~/.profile" ad, "export PATH=/user/local/bin:/usr/local/sbin: $PATH"

Step 5: Install Python language interpreter by typing "Brew install Python".

Step 6: Wait for the installation process to complete.

Once you have installed Python, you can use your coding application to work on programming Markov Models using Python.

## Programming a Hidden Markov Model

Let us take a look at the different components we will be using.

Import numpy as np

```
Import pandas as pd
Import netwrokx as nx
Import matplotlib.pyplot as plt
%matplotlib inline
```

Next, we will need to input all the information we have on the Markovian model being considered.

The variables that we need to know are the states, the probabilities of transition from one state to the other (A) and the emission probabilities (B), the initial probability distribution P an any other information that can be found. Let us take the example of a baby. There would be three states that the baby would normally be in "sleep", "eat" and "cry". This can be represented as follows:

S = {sleep, eat, cry}

We would then need to look at the probabilities of the states and also look at the initial probability distribution P. Let us assume that there is a 40% chance of the baby sleeping, 35% chance of the baby crying and 25% chance of the baby eating. This would mean:

P = {0.40, 0.35, 0.25}

This information is represented as follows in Python:

States = ['sleeping', 'eating', ' crying']

P= [0.40, 0.25, 0.35]

State_space = pd.Series(pi, index = states, name = 'states')

Print(state_space) #ensure that the states and their probabilities are matched correctly
Print(state_space.sum()) #ensure that the probabilities add up to 1

You have to identify some output that would match each state correctly and also provide you with a distribution where the probability sum up to 1. You have to ensure that you do not use any symbolic notations in Python since the language would not accept those.

Let us now input the state transition probabilities. Let us assume that the probabilities are as follows:

Sleeping -> Sleeping = 0.40
Sleeping -> Eating = 0.20
Sleeping -> Crying = 0.40
Eating -> Sleeping = 0.45
Eating -> Eating = 0.45
Eating -> Crying = 0.10
Crying -> Sleeping = 0.45
Crying -> Eating = 0.25
Crying -> Crying = 0.30

*Note: '->' indicates moving from one state to the other.*

This information would need to be inputted into Python in the following manner:

O_df = pd.DataFrame (columns = states, index = states)
O_df.loc [states[0]] = [0.4, 0.2, 0.4]
O_df.loc [states[1]] = [0.45, 0.45, 0.1]
O_df.loc [states[2]] = [0.45, 0.25, 0.3]
Print (O_df)
O=O_df.values #ensure that the matrix has been created correctly and the values are all placed correctly

Print ('\n',O,O.shape,'\n\) #ensure that the shape of the matrix is right
Print (O_df.sum(axis=1)) #ensure that the probabilities add up to 1

Printing this will show the matrix to you and will give you the exact shape of the matrix (3x3). The sum of the individual probabilities will help you identify if the total is 1 or lesser than 1. Make sure that you change the probabilities appropriately if there is any discrepancy.

In order to make this a Hidden Markov Model, you would need to recognize certain factors of the baby which would make him move into one of the states mentioned. For instance, if the baby were sick, he or she would probably wail throughout the night. If he or she were completely healthy, it would be a whole other state. There will need to be a set of probabilities that would need to be created in order to take these states into account – since it is really likely for the baby to be healthy or sick.

Hidden_states = ['healthy', ' sick'0
P = [0.5,0.5]
State_space= pd.Series (P, index = hidden_states, name= 'states')
Print (state_space) #make sure that the hidden states and probabilities have been matched correctly

Print ('\n', state_space.sum()) #ensure that the probabilities are equal to 1

We will need to print one more time to ensure that the states and probabilities have been matched perfectly and that the total probability is equal to 1.

We would now need to input the emission probabilities.

Observable_states = states
B_df = pd.DataFrame (columns = observable_states, index = hidden_states)
B_df.loc[hidden_states[0]] = [0.2,0.6,0.2]
B_df.loc[hidden_states[1]] = [0.4,0.1,0.5]
Print(B_df)
B=B_df.values  #ensure that the values correspond to the states
Print ('\n',B,B.shape,'\n') #ensure that the emission probabilities add up to 1

This matrix should be 2x3, which represents the two hidden states and the three observed states.

You have successfully created a Markov Model for a baby. You can use a similar program to create another Markov Model for the weather and verify if you have understood the concepts well.

# **Conclusion**

I want to thank you once again for choosing this book.

You should now have a better understanding of the Markov Model and also learn to program the basic Markov Model on Python.

To summarize, we have looked at the axioms that one would need to be clear with when learning Markov Models. These included a number of axioms that are integral to Markov Models. We then looked at Markov Models and Hidden Markov Models and covered some aspects of Markov Models. We then looked at Data Mining and also tried to understand different aspects of Data Mining. We answered the question of how Data Mining is used and also tried to see why it is used.

We then covered machine learning and looked at Supervised and Unsupervised Machine learning in great detail. We looked at the different ways in which machine learning is used and also its applications. We also took a look at artificial intelligence and how it is important in big data analysis. This chapter also covered certain reasoning skills used by AI Systems to understand data and come to a conclusion or predict the future. We then looked at how Data Science plays a key role in Artificial Intelligence.

The last section of the book deals with creating a Markov Model using Python. It has been explained very clearly in the book how a user can download and use Python. A sample code has been provided to you in the chapter, which will help you fit a three state Markov Model. However, you have to remember that this model is only code and it can be tweaked for your purposes.

By now, you should feel comfortable with the different terms in machine learning and also understand the mathematical aspects that lie behind it. I hope you feel more informed about the coding and mathematics that drives a great deal of data that is collected and used in the data science and artificial intelligence industries. I hope you are inspired to learn more.

If you have enjoyed reading this book please leave a quality review on Amazon.com Thank you! :) Please see link below..

http://amzn.to/2ey04bI

www.ingramcontent.com/pod-product-compliance
Lightning Source LLC
Chambersburg PA
CBHW070257230526
45470CB00002B/623